Active Minds

ZOOM INTO SPACE

THE MOON

our neighbor in space

Ellen Lawrence

sequoia kids media

Photography © NASA, Ruby Tuesday Books, Shutterstock 2021 Adelart; Amguy; Balabolka; Dotted Yeti; ideyweb; JLStock; Marina Yesina

Consultants:

Suzy Gazlay, MA
Recipient, Presidential Award for Excellence in Science Teaching

Kevin Yates
Fellow of the Royal Astronomical Society

Published by Sequoia Kids Media, an imprint of Sequoia Publishing & Media, LLC

Sequoia Publishing & Media, LLC., a division of Phoenix International Publications, Inc.

8501 West Higgins Road
Chicago, Illinois 60631

Customer Service: cs@sequoiakidsbooks.com

www.sequoiakidsmedia.com

ISBN 978-1-64996-209-6

active minds

ZOOM INTO SPACE

THE MOON

table of contents

Words shown in **bold** in the text are explained in the glossary.

Welcome to THE MOON

Imagine a world that is thousands and thousands of miles from Earth.

The ground is covered with rocks and thick, gray dust.

This place is home to tall mountains, flat **plains**, and gigantic holes called **craters**.

During the day, it is scorching hot.

At night, it is much colder than the coldest place on Earth.

Welcome to the Moon!

This photo shows an **astronaut** named Charles Duke standing on the Moon. Other than Earth, the Moon is the only place in space where humans have ever stood.

Space Neighbors

The Moon and our home **planet** Earth are traveling through space together.

As they move through space, the Moon is **orbiting**, or circling, Earth.

It takes the Moon just over 27 days to orbit Earth once.

When we see the Moon in the night sky, it looks white or grayish-white.

That's because the Sun is shining on the Moon and lighting it up.

From Earth, it looks as if the Moon is changing shape. It doesn't really, though. As it travels around Earth, we see different parts of its bright, shining surface.

These photos show the different ways that we see the Moon from Earth.

The Solar System

Earth and the Moon are traveling in a big circle around the Sun.

Earth is one of eight planets circling the Sun.

The planets are called Mercury, Venus, Earth, Mars, Jupiter, Saturn, Uranus, and Neptune.

Some of the other planets have **moons** as neighbors, too.

In addition to planets and moons, rocky **asteroids** and icy **comets** orbit the Sun.

Together, the Sun and its family of space objects are called the solar system.

Most of the asteroids orbiting the Sun are in a ring called the asteroid belt.

An asteroid

The solar system

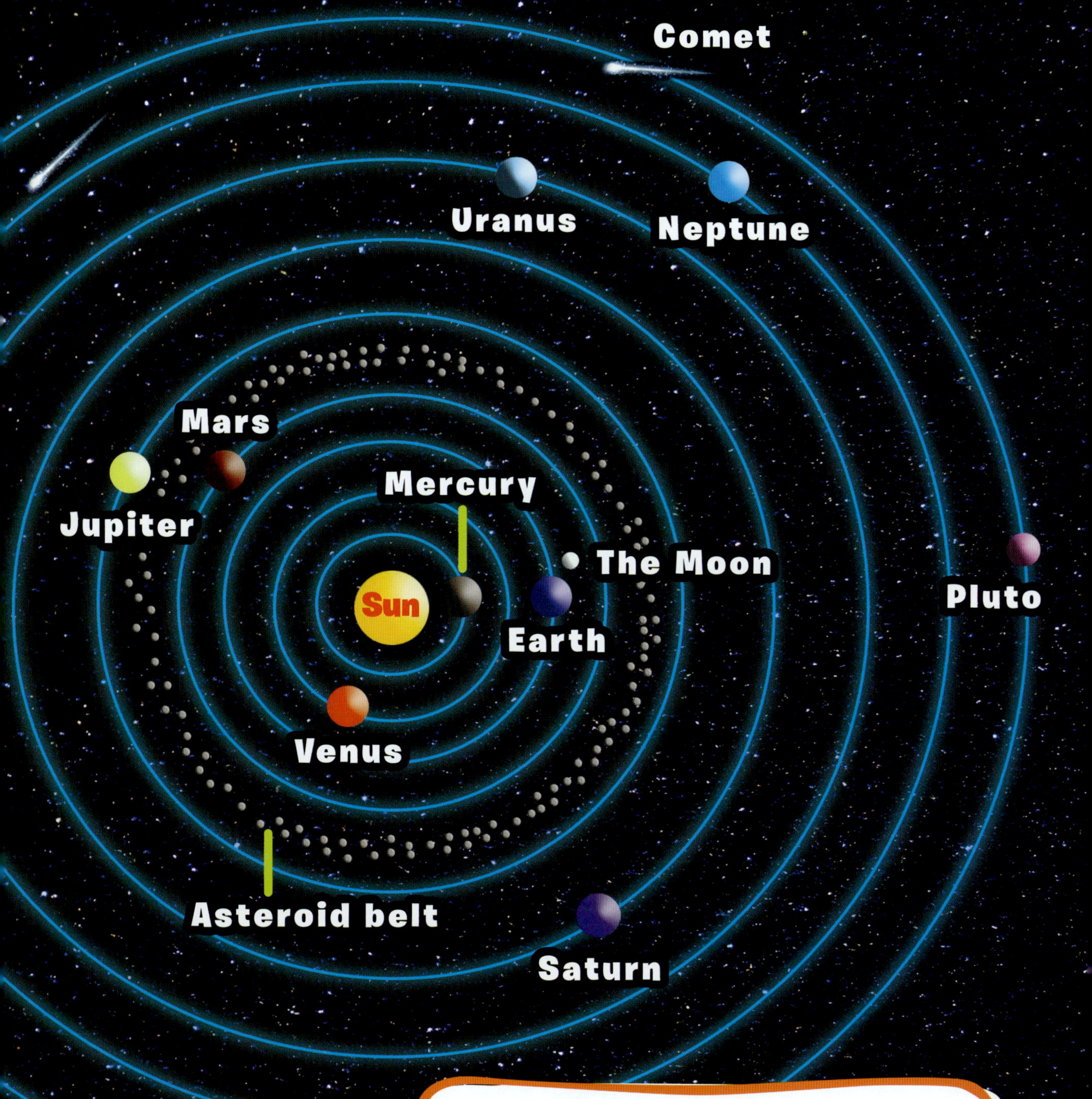

The solar system is home to small planets, called **dwarf planets**. Pluto is a dwarf planet.

A Closer Look at The Moon

The Moon is just over 2,000 miles (3,000 km) wide.

It is made mostly of rock with a ball of iron in the center.

Earth is covered with a thick layer of **gases** called an **atmosphere**.

It's these gases that make Earth's sky look blue.

The Moon does not have an atmosphere.

From the Moon, the sky looks black whether it is day or night.

How Big is the Moon?

Earth's atmosphere contains **oxygen** that humans and other animals need to breathe. Because the Moon has no atmosphere, there is no oxygen to breathe on the Moon.

The Moon Up Close

If you look up at the Moon, you can see large dark areas on its surface.

Hundreds of years ago, people thought these dark areas might be oceans.

Today, we know they are wide, flat plains made of dark, smooth rock.

It's also possible to see mountains on the Moon through a telescope.

Some of the Moon's mountains are up to 3 miles (5 km) tall.

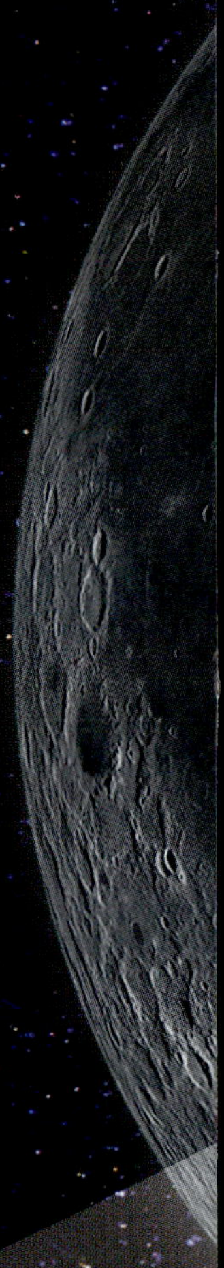

The surface of the Moon is covered with a thick layer of dust and rocks. Some of the rocks are as big as trucks!

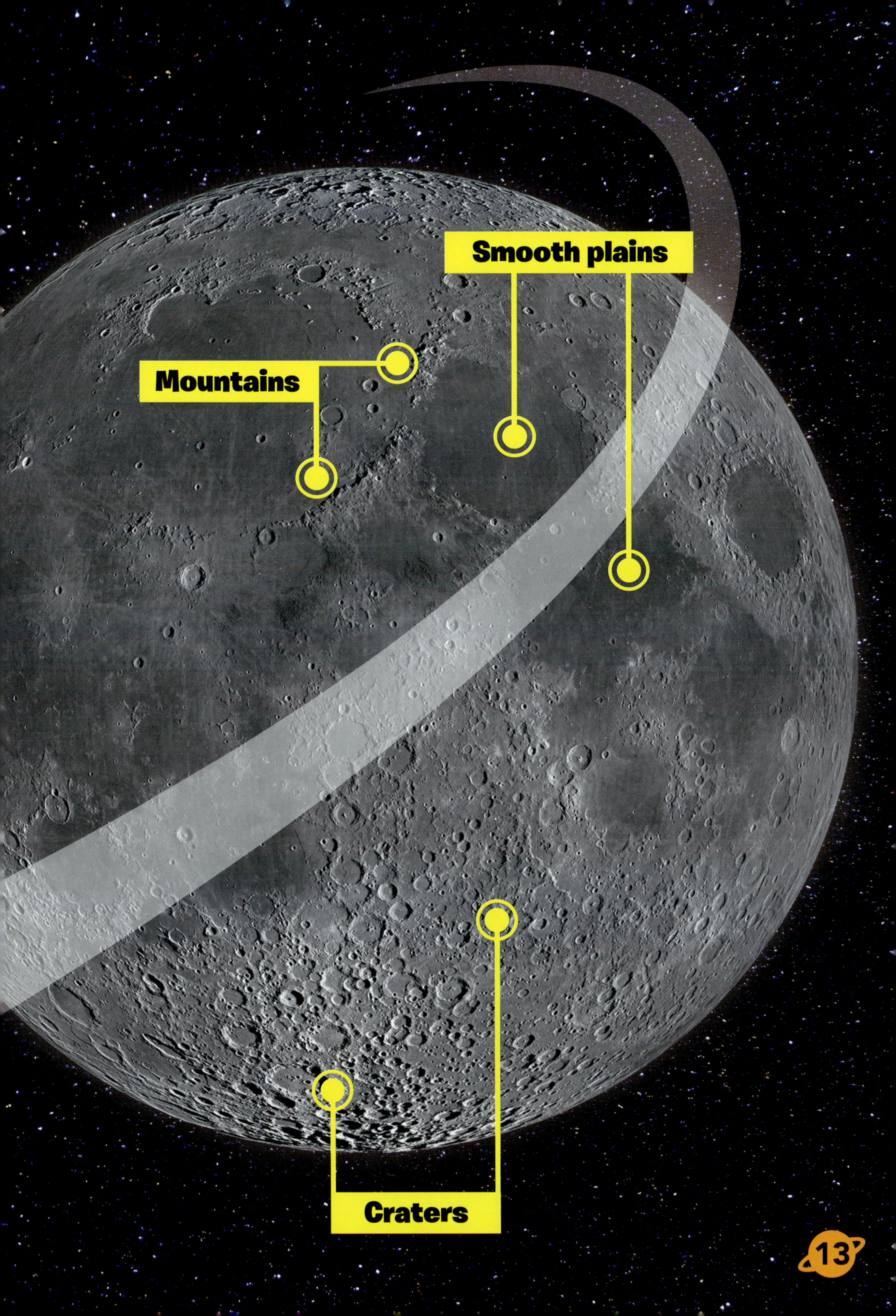
Smooth plains
Mountains
Craters

A Gigantic Crater

The surface of the Moon is covered with hundreds of thousands of craters.

These craters are made by objects such as asteroids and comets that hit the Moon.

Many of the craters are hundreds of miles wide.

The biggest craters are known as **impact basins**. The largest impact basin on the Moon is 1,600 miles (2,600 km) across.

If this gigantic crater was on Earth, it could cover half of the United States!

The Moon's gigantic impact basin is called the South Pole-Aitken basin. It is at the Moon's south pole. The basin is over 8 miles (13 km) deep.

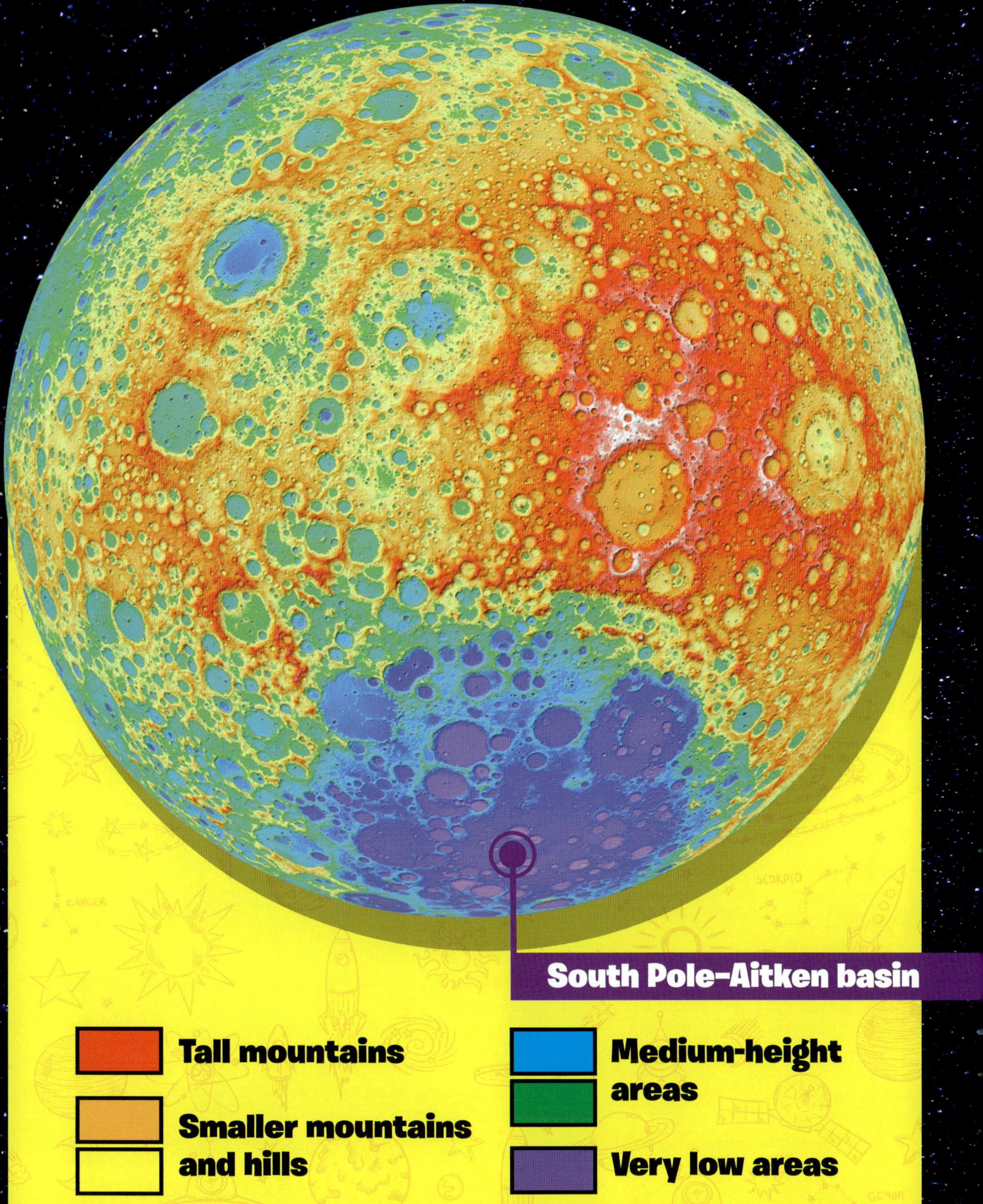

This picture of the Moon was created on a computer. The colors show the different heights of the land.

walking on The Moon

In July 1969, three astronauts left Earth aboard a spacecraft named *Columbia*.

When they reached the Moon, astronaut Michael Collins stayed aboard *Columbia*.

Neil Armstrong and Buzz Aldrin flew down to the Moon's surface in a spacecraft called the *Eagle*.

On July 20, 1969, Armstrong and Aldrin became the first people to walk on the Moon!

The astronauts explored the Moon for two hours and 36 minutes.

Then they flew the *Eagle* back to *Columbia*, and headed home.

The *Eagle* lander

During their moonwalk, the astronauts collected Moon rocks. They also took photos and filmed the Moon with TV cameras.

There was oxygen inside the astronauts' helmets so they could breathe. Their spacesuits protected them from the extreme hot and cold temperatures on the Moon.

Studying the Moon

Since the first moonwalk, other astronauts and spacecraft have visited the Moon.

In 2009, a space **probe** called *LCROSS* blasted a rocket into a deep crater on the Moon.

The crash made a huge cloud of dust and other material rise up into space.

LCROSS studied the cloud and discovered tiny pieces of ice mixed in with the rocky dust.

This means there is frozen water inside craters on the Moon. If there is water on the Moon, people might one day be able to live there!

Many people have dreamed of building a moon base, or city, on the Moon. Inside the base's buildings, people would be protected from the heat and cold. There would also be oxygen for people to breathe.

An illustration of the *LCROSS* mission

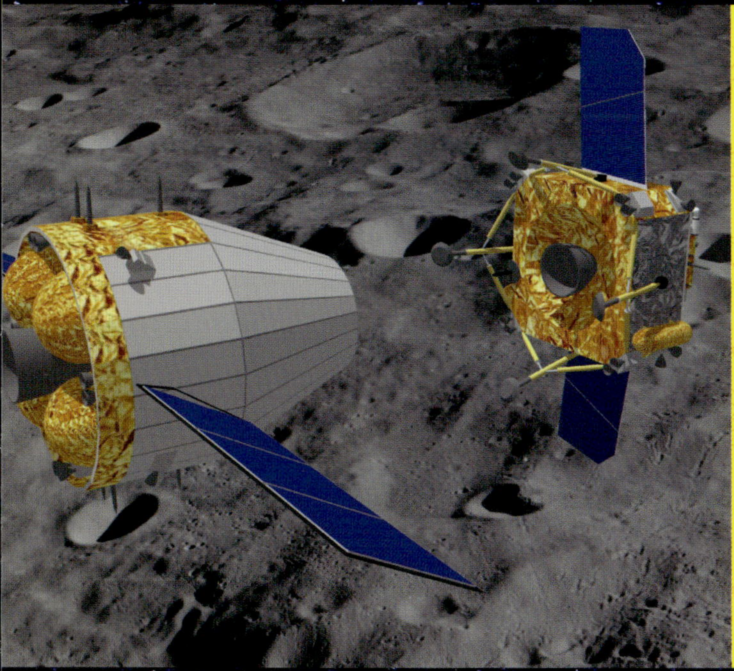

On December 1, 2020, China's Chang'e 5 mission sent a spacecraft to the Moon. A small lander visited the Moon's surface and collected rocks for scientists back on Earth to study. Then the spacecraft returned to Earth on December 16, 2020.

The Moon
Fact File

Here are some key facts about Earth's nearest space neighbor, the Moon.

How the Moon got its name

The word "moon" comes from very old words meaning "month" and "measure." The Moon's orbit around Earth has been used to measure the number of days in a month.

Astronauts on the Moon

Twelve astronauts have walked on the Moon.

The solar system's moons

There are at least 180 moons orbiting planets in the solar system. This picture shows the sizes of the five biggest moons compared to Earth's moon.

The Moon Earth

Ganymede Jupiter

Titan Saturn

Callisto Jupiter

Io Jupiter

Triton Neptune

The Moon's size

2,159 miles (3,475 km) across

Length of the Moon's orbit around Earth

1,499,618.58 miles
(2,413,402.16 km)

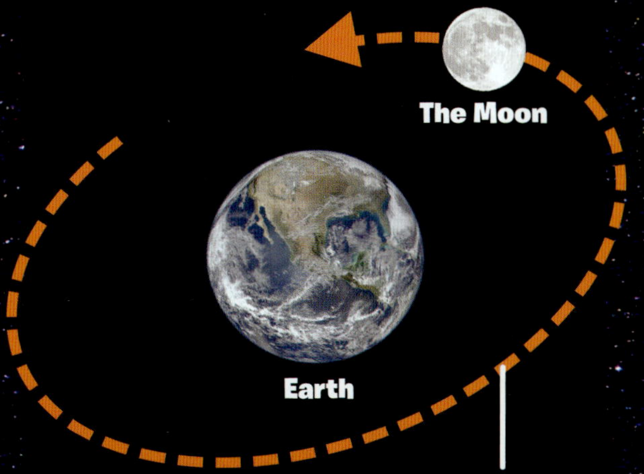

The Moon's distance from Earth

The closest the Moon gets to Earth is 225,623 miles (363,104 km).

The farthest the Moon gets from Earth is 252,088 miles (405,696 km).

Temperature on the Moon

Highest: 253°F (123°C)
Lowest: -387°F (-233°C)

Average speed at which the Moon orbits the Earth

2,287 miles per hour
(3,681 km/h)

Get Crafty

Make Moon craters

Rocky space objects hit the surface of the Moon and make craters. Using pebbles and plaster of Paris, you can make a model of the Moon's craters.

You will need:

- A bowl
- Plaster of Paris
- Water
- A spoon
- A foil pie tin
- Newspaper
- Pebbles or small pieces of rock

1. Put the plaster of Paris into the bowl. Add some water and stir until the mix looks like smooth pancake batter. Keep adding plaster or water until the mix is right. You will need enough mix to fill the pie tin.

2. Pour the mix into the pie tin and place the tin on the floor on top of some newspaper. Now, wait for the mix to start setting. Keep touching the mix with your finger. When it feels like ice-cream that is just getting soft, it's time to make craters!

3. Drop a pebble into the pie tin. It will make a crater in the plaster mix! Quickly and carefully remove the pebble, then try again with a different pebble.

4. When the plaster mix is covered with craters, leave the plaster to set hard. This should take about 30 minutes. Then carefully remove your model from the pie tin.

Glossary

asteroid (AS-teh-royd) A large rock that is orbiting the Sun. An asteroid can be as small as a car or bigger than a mountain.

astronaut (AS-troh-nawt) A person who is trained to go into space in a spacecraft.

atmosphere (AT-muh-sfeer) A layer of gases around a planet, moon, or star.

comet (KAH-mit) A space object made of ice, rock, and dust that is orbiting the Sun.

crater (KRAY-tur) A bowl-shaped hole in the ground. Craters are often caused by asteroids and other large, rocky objects hitting the surface of a planet or moon.

dwarf planet (DWARF PLAN-et) A round object in space that is orbiting the Sun. Dwarf planets are much smaller than the eight main planets.

gas (GASS) A substance, such as oxygen or helium, that does not have a definite shape or size.

impact basin (IM-pact BAY-sin) A very large crater that can be hundreds of miles wide.

moon (MOON) An object in space that is orbiting a planet. Moons are usually made of rock, or rock and ice. Some are just a few miles wide. Others are hundreds of miles wide. Earth has one moon. We simply call it "the Moon."

orbit (OR-bit) To circle, or move around, another object.

oxygen (OX-ih-jin) An invisible gas in the air that you and other living things need to breathe.

plain (PLANE) A large, flat area of ground.

planet (PLAN-et) A large object in space that is orbiting the Sun. Some planets, such as Earth, are made of rock. Others, such as Jupiter, are made of gases and liquids.

probe (PROBE) A spacecraft that does not have any people aboard. Probes are usually sent to planets or other objects in space to take photographs and collect information. They are controlled by scientists on Earth.

Index

Read More

The Moon
Space: Facts and Figures
Nancy Dickmann
Windmill Books
2019

Mysteries of the Moon
Space Mysteries
Rebecca E. Hirsch
Lerner Publishing Group
2021

Bonus Downloadable Activities!

www.sequoiakidsmedia.com

ARIES
VIRGO
SAGITTARIUS
LEO
TAURUS
CAPRICORN
SCORPIO
PISCES
ARIES
AQUARIUS
VIRGO
GEMINI
LIBRA
TAURUS
CAPRICORN
CANCER
PISCES
ARIES